William App Jones

A Contribution to the Knowledge of Dicarbonyl Cuprous Chloride

William App Jones

A Contribution to the Knowledge of Dicarbonyl Cuprous Chloride

ISBN/EAN: 9783337222079

Printed in Europe, USA, Canada, Australia, Japan

Cover: Foto ©berggeist007 / pixelio.de

More available books at **www.hansebooks.com**

A Contribution to the Knowledge

of

Dicarbonyl Cuprous Chloride

Dissertation

Submitted to the Board of University Studies of the Johns Hopkins University

by

William Apo Jones

1898

Contents

Introduction	1
Preparation of material	
1. Cuprous Chloride	7
2. Carbon monoxide	10
Preparation of dicarbonyl cuprous chloride	11
Analysis of the compound	
1. Decomposition method	15
2. Absorption method	25
Behavior under reduced pressure	
1. Compound in dilute hydrochloric acid	36
2. Compound in water	43
3. Compound in concentrated acid	49
Absorption of carbon monoxide by cuprous chloride in pyridine	52
Behavior of this compound under increased temperature	59
Action of oxygen on dicarbonyl cuprous chloride	66
Action of chlorine	73

III

Action of hydrogen 81
Action of nitrogen 83
Action of acetylene on carbon monoxide ... 83
        Discussion of Results.
Action of oxygen on dicarbonyl cuprous chloride ... 88
The constitution and chemical nature
   of dicarbonyl cuprous chloride ... 98
Evidence for and against molecular compounds ... 101
Effect of pressure changes on the
   molecular weight of compounds ... 120
The disruptive influence of reduced
   pressure on chemical combinations ... 124
Power of increased pressure to
   retard decomposition ... 126
Effect of an excess of one constituent ... 129
Effect of a solvent ... 131
Inorganic compounds containing carbon monoxide ... 148
Conclusions ... 154
Bibliographica ... 151

Acknowledgment

The present work was undertaken at the suggestion of Prof. Remsen and his constant interest and advice have been of the greatest value.

I wish to thank him most sincerely for instructions received both in the lecture-room and laboratory; I feel especially indebted to him for the high ideals, not of chemistry alone but of life as well, which personal contact with him cannot fail to inspire.

My thanks are also due Profs. Morse and Clark and Dr. Mathews for the instruction received at their hands.

A Contribution to the Knowledge of Dicarbonyl Cuprous Chloride

Introduction

In attempting to determine the free oxygen in illuminating gas by means of an ammoniacal solution of cuprous chloride, Leblanc Stas and Doyèn[1] observed that very large amounts of carbon monoxide and ethylene were also absorbed.

Leblanc then made an investigation as to the absorption of carbon monoxide by cuprous chloride and found that carbon monoxide is absorbed by a

---

[1] Compt. rend. (1850) 30, 483

hydrochloric acid solution of cuprous chloride with about the same facility that carbon dioxide is absorbed by caustic potash; however, in the former instance the temperature of the solution is only very slightly elevated. He also observed that an ammoniacal solution of cuprous chloride absorbs carbon monoxide, in the absence of air, to exactly the same extent as does the hydrochloric acid solution, the proportion of carbon monoxide to cuprous chloride being, as he thought, 1:1 in both cases.

He also found that the hydrochloric acid solution of cuprous chloride, saturated with carbon monoxide, could be diluted with water to a considerable extent without any precipitate being formed, just as before the absorption, and by the dilution no evolution of gas is caused.

The same was found true for alcohol though ether seems to decompose the compound to some extent.

Urbain also noticed that by elevating the temperature or removing the atmospheric pressure, the carbon monoxide is given off as such from the solution. He did not succeed in isolating a definite compound and performed no other experiments with the solution.

He did discover, however, that ammoniacal solutions of the cuprous salts in general possess the ability to absorb carbon monoxide.

He thought the absorption of carbon monoxide by cuprous chloride was to be compared to the absorption of nitric oxide by ferrous sulphate.

It was especially emphasized that this absorption might furnish evidence in support of the radical theory, which occupied the minds of chemists so generally at that time. Since cyanogen also is absorbed by a hydrochloric acid solution of cuprous chloride, he reasoned that in the chlorformic ethers ($CO\genfrac{}{}{0pt}{}{Cl}{OR}$) carbonyl chloride ($CO\genfrac{}{}{0pt}{}{Cl}{Cl}$) etc carbon monoxide plays the part of a radical, just as does cyanogen.

cyanogen chloride (CN.Cl).

In 1856 Berthelot[1] succeeded in isolating the compound formed by the action of carbon monoxide on cuprous chloride in hydrochloric acid solution. His method of preparation consisted in passing a stream of carbon monoxide through the cold saturated solution of cuprous chloride, which was repeatedly shaken to facilitate absorption. In this way he obtained beautiful pearly crystals, which altered very rapidly in the air, making it impossible to prepare them for analysis without at least some loss of carbon monoxide. The figures obtained by his analyses approximated those required by the formula

$4Cu_2Cl_2 \cdot 3CO \cdot 7H_2O$, but since the compound alters so rapidly, he believed it to be $Cu_2Cl_2 \cdot CO \cdot 2H_2O$, and this is the composition usually ascribed to the compound, when any at all is given.

Evidently this formula is based upon very unsatisfactory evidence and the present work was undertaken to determine more accurately, if possible, the composition of the substance, to study its conduct under different conditions of temperature and pressure, and to determine the chemical activity of the carbon monoxide at the moment of its liberation.

Preparation of Material.
Cuprous Chloride.

This was prepared by the action of aqua regia on copper turnings - the latter being in excess. In its preparation about half a liter of concentrated hydrochloric acid was poured into a flask containing about 300-350 grams of copper turnings, and the acid heated almost to boiling; then concentrated nitric acid was added from time to time, only small amounts being used in order that the solution might not boil over. After being heated for about half an hour, the hot concentrated solution was quickly filtered through an asbestos filter, and to the filtrate cold water was added, the solution being stirred constantly. Four or five liters of water must be added to one liter of the acid solution to cause complete precipitation of the cuprous chloride. Obtained in this way it is a fine, perfectly white powder, and can be easily filtered off and washed free from acid; but in order that the material may not be oxidized by contact with the air, it is necessary to keep enough water in the funnel to cover the cuprous chloride. The final washings were made with glacial acetic acid, and the material dried in an air-bath, at a temperature of 115°-125°C, for 12 hours.

Proceeding in this way, it is possible to obtain the material for fectly dry and without any alteration in color, and a specimen kept for more than six months in a glass-stoppered bottle was still perfectly white. However, it must be dried immediately after washing with acetic acid + the heating must be continued for several hours after the odor of the acid has disappeared. Otherwise it will invariably turn green in a short while, no matter how carefully it may be stoppered.

In powdering the material it is well to protect the mouth and nose, as the finely divided cuprous chloride acts as a painful irritant to the mucous membranes.

The analysis of the material gave the following results:

I   0.2012 grams material gave 0.1607 g. CuO
II 0.2112 grams material gave 0.1692 g. CuO

Calculated for
Cu_2Cl_2 Found
Cu = 64.06 %   I. 63.72 %
                          II. 63.94 %

Preparation of Carbon Monoxide.

The carbon monoxide used in almost all the experiments was prepared by the action of concentrated sulphuric acid on potassium ferrocyanide. The gas was passed through four or five wash-bottles containing concentrated potash solution until on testing the gas it was found to contain practically no air, and was then collected in a glass gasometer.

In all the calculations from here on the following atomic weights have been used:

...made slightly alkaline. The carbon monoxide obtained in this way was entirely free from carbon dioxide and sulphur dioxide, and contained not more than 0.2 of 1% of air, and this slight amount of air came no doubt from that which was dissolved in the water used in the gasometer.

Preparation of Dicarbonyl Cuprous Chloride

In all cases the same general method was employed. The cuprous chloride was placed in a small balloon flask (D), which in turn was immersed to a given point in ice water. The flask was tightly fitted with a three-hole rubber stopper, through one opening was passed a small dropping-funnel (B) and through the other two glass tubes bent at right angles - one of which was connected with a gas burette (A), the other (C) with a suction-pump.

After the introduction of the cuprous chloride and the immersion of the flask in ice-water, connection was ~~was~~ made with the pump and the flask exhausted - the tube leading to the gas burette having been closed. After the exhaustion the stop-cock C was closed + enough hydrochloric acid introduced through the dropping-funnel to make a thin paste with the cuprous chloride. The tube leading to the gas burette, which contained the carbon monoxce, was then opened and the absorption flask alternately shaken and immersed in ice-water.

At first the absorption is quite rapid, but toward the end of the operation it becomes very slow and constant shaking is necessary; even then it is only after several hours that the absorption is complete. If at any time the paste in the flask became too thick to be shaken easily, more hydrochloric acid was introduced; not in excess, however, for both the cuprous chloride and dicarbonyl cuprous chloride are soluble in an excess of the acid, and I found it almost impossible to get the latter out of this solution in good yields.

By the above process it is

possible to obtain the compound in beautiful pearly plates, — the crystals seemingly being perfectly pure — and in any quantity desired. By adding ice-water to the solution the crystals can be filtered and washed free from acid, and after some experience they can be fairly well dried between sheets of drying-paper; however, the paper should previously be cooled and the pressure should be applied with a previously cooled spatula. Exposure of the crystals to the air, at the room temperature, for more than a minute or two, causes the entire mass to become intensely green and carbon monoxide is given off.

Analysis of the Comp......
I Decomposition Method

From what has been said it is evidently impossible to dry and then weigh the compound for analysis. In the first instance it was attempted to dry the compound as quickly and thoroughly as possible, every precaution being taken to prevent decomposition, and then to decompose the compound by elevating the temperature, the products being either measured or weighed.

The apparatus consisted of a small bulb (B), of about 6 c.c. capacity, into which the compound was introduced for the decomposition. This was provided with a tight-fitting, two-hole rubber stopper. Through one opening was passed a small glass tube, bent at right angles, which extended to the bottom of the flask and was connected with a calcium-chloride tube. This was closed during the decomposition. Through the other opening was passed a similar glass tube which extended just through the stopper and was connected with a calcium-chloride tube (C), to be used for collecting the water given off from the compound by its decomposition. This tube was protected by another calcium-chloride tube (D), the latter being connected with a gas-measuring tube (E) in which the ~~the~~ carbon monoxide was to be col lected and measure. During the decompositions the little bulb containing the material for analysis, (B), was immersed to a given mark in a water bath maintained at 80°C, and blank determinations were previously made to ascertain the correction to be applied to the volume of the carbon monoxide collected, the correction being the expansion of the gas in the bulb.

In making the determinations the compound was prepared as previously described, washed free from hydrochloric acid by means of ice-water and quickly transferred from the Witt plate, on which it was washed, to drying-paper, pressed between folds of the paper with a cold spatula x transferred from one [paper] to another, until on pressing with a cold spatula a moist spot [no longer] is made on the drying-pan.

This operation [does] if not require more than one minute, and none but cold objects should touch the crystals, otherwise they immediately become green. Just as soon as dry, the crystals were quickly placed in the decomposition-bulb (B), which had been previously weighed. The stopper fitted in tightly and connection made with the weighed calcium-chloride tube (C), which was to retain the water given off by the compound; the connections with the second calcium-chloride tube (D) and the gas-measuring tube (E) were opened at the same ture. The decomposition-bulb was then immersed to the mark in the water-bath and a temperature of 80°C maintained until there was no further change in the volume of gas in the measuring-tube (E), and this volume was noted, corrected for temperature, pressure, and water vapor, the correction due to the expansion of the gas in the decomposition-bulb being also applied. In this way the volume of carbon monoxide given off from the compound was ascertained.

Of course only a small portion of the carbon monoxide given off ever reaches the measuring-tube, but the air displaced was measured; accordingly can was expressed to have the pressure in all parts of the apparatus the same and equal to the atmospheric pressure, before the final readings of the gas volume were made.

The volume of the carbon monoxide having been ascertained, the measuring-tube was disconnected and dry carbon dioxide passed through the calcium-chloride tube A, and through the decomposition-flask, the latter remaining in the water-bath at 80°C. In this way the material analyzed could be thoroughly dried and the current of gas was continued until there was no further change in the weight of the bulb. Then the bulb and weighed calcium-chloride tube were again carefully weighed, the difference between the first and last weighings being the weight of cuprous chloride and carbon monoxide respectively. The total weight of the constituents, obtained by the decomposition of the compound, being known, the percentage of each constituent was calculated.

The cuprous chloride remaining in the decomposition-bulb was always found to be slightly green in color. This seemed due not to the action of the carbon dioxide, but to oxidation by the air in the bulb when the compound was introduced. In order to obviate this difficulty a bulb as small as possible was employed and by analysis the cuprous chloride was found to be within about 0.5 of 1% pure.

While this method is theoretically very simple, it is in practice almost impossible to obtain concordant results by means of it. The great difficulty is the extreme instability of the compound in contact with the air, making it almost, if not quite, impossible to dry it without loss of carbon monoxide and oxidation of the cuprous chloride. A large number of analyses were made in this way, but the results were not satisfactory, even after considerable skill had been acquired in drying and handling the compound, and the method was finally given up for a still simpler one

However this was the only method used by means of which it was possible to determine the percentage of water, and while I would place rather small faith in this method alone, yet, supported as it is by the more accurate method, the results do have a certain value — a value that can enable one to ascertain with a fair degree of accuracy the amount of water in the compound.

Below I give the results of three analyses, fair specimens of the best made, though others showed even greater variations than these.

Determination No. I.

Weight Cu_2Cl_2 = 0.3280 gr. = 60.78%
" H_2O = 0.1220   = 22.61%
" CO = 0.0896   = 16.60%
Total   = 0.5396   99.99

Or calculated only for cuprous chloride and carbon monoxide, we have

(1) Weight Cu_2Cl_2 = 0.3280 gr.   = 78.54 %
   " CO = 0.0896   = 21.45 "
   Total   0.4176   99.99

Determinat⁓ No. II.

Weight Cu_2Cl_2 = 0.2180 gr.   = 62.21 %
   " H_2O = 0.0740   = 21.11 "
   " CO = 0.0584   = 16.67 "
   Total = 0.3504   99.99

Calculated in terms of Cu_2Cl_2 and CO.

Weight Cu_2Cl_2 = 0.2180 gr.   = 78.87 %
   " CO = 0.0584   = 21.13 "
   Total = 0.2764   100.00

Determination No III.

Weight Cu_2Cl_2 = 0.2170 gr   62.41 %
" H_2O = 0.0750   21.57 %
" CO = 0.0557   16.02 %
Total   = 0.3477   100.00

Calculated in terms of Cu_2Cl_2 and CO.
Weight Cu_2Cl_2 = 0.2170 gr   = 79.57 %
" CO = 0.0557   = 20.42 %
Total   = 0.2727   99.99

II Absorption Method.

Finding that it was impossible to obtain sufficiently accurate results by the method just described, it was decided to take a known quantity of cuprous chloride, and determine the amount of carbon monoxide it would absorb. The apparatus used w same as that employed in the preparation of the compound and described on page 11. From one to two grams of cuprous chloride was carefully weighed and introduced into the absorption-bulb (II), which was then closed, immersed to the mark in ice-water and exhausted. The exhaustion being complete, the connection with the pump was closed, enough concentrated hydrochloric acid introduced to form a thin paste with the cuprous chloride, the connection with the gas-measuring tube opened and carbon monoxide allowed to enter, the volume of gas in the tube being quickly noted as soon as atmospheric pressure was reached in the absorption-bulb. The bulb was then alternately shaken in dipped in ice-water until no more carbon monoxide could be absorbed. This required from two to three hours. Toward the end the absorption is extremely slow, and almost constant shaking is required so that the unchanged cuprous chloride may be exposed. Also care must be exercised that the material does not stick to the sides of the bulb, else some of the unchanged cuprous chloride will be covered by the dicarbonyl cuprous chloride, and so complete absorption be prevented. It is needless to say that the absorption was made as complete as it was possible to get it. In some cases the absorption-bulb was kept in ice-water for several days, being shaken at intervals, but no was absorbed after the first three hours, when this part of the operation had received the proper treatment.

It was only after a month's work, in which time at least fifty determinations were made, that I could obtain concordant results, due most likely to incomplete absorption. All previous determinations were then discarded and a new series of about twenty five made, in which the results varied by only a few tenths of one per cent.

Determinations were also made to see if the strength of acid used had any influence on the amount of carbon monoxide absorbed. It was found to have no effect at all, only that the absorption was mor rapid the more concentrated the acid used — due, no doubt, to the greater solubility of both compounds in the stronger acid.

These results led me to try pure water instead of hydrochloric acid, and while the absorption is very slow, yet exactly the same amount of carbon monoxyde is absorbed as when the acid is used. This fact has never been noticed before, and since the same amount of gas is absorbed and the appearance of the crystals is identical in all cases, it tends to show that where hydrochloric acid is used as a solvent none of the acid enters into the composition of the compound.

When the bulb containing carbonyl cuprous chloride is heated to 60° or 70°C, exactly the same amount of carbon monoxide is given off as was absorbed in the first instance; on cooling the bulb it is exactly reabsorbed and this operation can seemingly be repeated indefinitely. Dry cuprous chloride does not absorb dry carbon monoxide.

I give below four analyses in the order in which they were made.

Deter. No IV.   Solvent H_2O
  Weight Cu_2Cl_2 = 2.00060 g.
  " CO absorbed = 0.54374 [1]
                              CO = 21.37%

Deter. No V.   Solvent 1 part HCl to 4 parts H_2O
  Weight Cu_2Cl_2 = 1.00450 g.
  " CO absorbed = 0.27031
                              CO = 21.20%

[1] In all cases when the weights or volumes of gases are employed, the corrections for temperature, pressure and water vapor have been applied.

Deter. No. VI.   Solvent 1 part HCl to 2 parts H₂O
Weight Cu_2Cl_2 = 1.00690 gr.
 " CO absorbed = 0.27332

$$CO = 21.35\%$$

Deter. No VII.   Solvent conc. HCl
Weight Cu_2Cl_2 = 1.00270 gr.
 " CO absorbed = 0.27489

$$CO = 21.5\%$$

I am aware that there is one considerable source of error in this method. The compound formed, as well as the cuprous chloride, is insoluble in water and only slightly soluble in hydrochloric acid, so that — no matter how long the absorption be continued nor how vigorously the bulb is shaken, an unchanged cuprous chloride in away or enclose in the dicarbonyl cuprous chloride formed and the results obtained in this way will invariably be slightly low.

As regards the formula expressed by these results - below are given the percentages of carbon monoxide required by different formulas.

$Cu_2Cl_2 \cdot CO$; $CO = 12.41\%$
$4 Cu_2Cl_2 \cdot 7CO$; $CO = 19.91\%$
$5 Cu_2Cl_2 \cdot 9CO$; $CO = 20.37\%$
$6 Cu_2Cl_2 \cdot 11CO$; $CO = 20.66\%$
$12 Cu_2Cl_2 \cdot 23CO$; $CO = 21.40\%$
$Cu_2Cl_2 \cdot 2CO$; $CO = 22.13\%$

It is evident that the formula considered probable by Leblanc $Cu_2Cl_2 \cdot CO$ and the one assigned to the compound by Berthelot ($Cu_2Cl_2 \cdot CO \cdot 2H_2O$) cannot be entertained since the percentage of carbon monoxide in $Cu_3Cl_2.CO$ is only 12.4%; the average in my determinations is about 21.4%.

Exactly 21.4% of carbon monoxide is required by the formula $12Cu_3Cl_2.23CO$; while such complex compounds are not unknown in inorganic chemistry, certainly they are quite rare, and while the percentage of carbon monoxide required for $Cu_3Cl_2.2CO$ is 22.13%, about 0.7 of 1% higher than the average found, at the same time it must be borne in mind that in both methods used there is a constant error, causing the percentage of carbon monoxide to be too low, so that it seems to me fair to accept the proportions $Cu_3Cl_2.2CO$ as highly probable

As regards the proportion of water contained in the compound we must rely on the "decomposition method". As stated on page 23, this method was found quite unsatisfactory, yet notice the rough agreement of the carbon monoxide as determined in this way. (CO = Deter. No I = 21.45%; No II = 21.13%; No. III = 20.72). These results show that while the compound has decomposed slightly, yet the average percentage of carbon monoxide (21.0%) is only slightly below the average obtained by the other method (21.4%). As previously stated, I think we are justified in assuming the proportions $BeCl_2 \cdot 2CO$ established. Keeping this part of the compound constant and varying the proportion of water, we obtain —

$Cu_2Cl_2 \cdot 2CO \cdot 3H_2O$; $H_2O = 17.55\%$
$Cu_2Cl_2 \cdot 2CO \cdot 4H_2O$; $H_2O = 22.11\%$
$Cu_2Cl_2 \cdot 2CO \cdot 5H_2O$; $H_2O = 26.19\%$

The average percentage of water as determined by the "decomposition method" (page 23) was seen to be 21.76%, being only 0.35% lower than required for $Cu_2Cl_2 \cdot 2CO \cdot 4H_2O$. On the other hand it is 4.21% higher than required by $Cu_2Cl_2 \cdot 2CO \cdot 3H_2O$, and 4.43% lower than required by $Cu_2Cl_2 \cdot 2CO \cdot 5H_2O$. Considering these facts, it seems highly probable that the composition of dicarbonyl cuprous chloride is expressed by the empirical formula $Cu_2Cl_2 \cdot 2CO \cdot 4H_2O$.

Behavior of Dicarbonyl Cuprous Chloride under Reduced Pressure

As was recognized by Leblanc, this compound loses carbon monoxide when the atmospheric pressure is removed. No exact statement has ever been made as to the pressure at which this decomposition begins, accordingly experiments were undertaken to ascertain just how the compound conducts itself when the pressure is removed.

In the preparation of the compound exactly the same precautions were observed as previously described, the apparatus used being modified in the following manner.

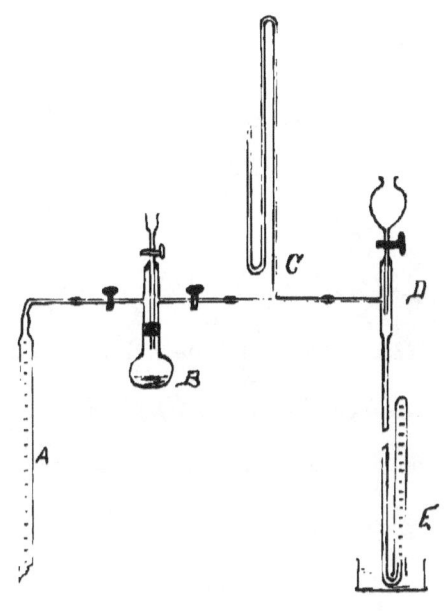

The compound was prepared in a small bulb of about 5cc capacity. This bulb was provided with a dropping funnel, of special design as shown, which could be disconnected or connected with a gas-measuring tube (A), and also a Sprengel pump. A manometer (C) was connected with the tube leading to the pump.

Blank experiments were made to ascertain the correction, on account of the expansion of the air in the bulb & tubes, which must be applied in the determinations. For this purpose the flask, containing the same amount of water used in the actual experiments, (30") was immersed to the mark in ice-water, and the pump started. The gas pumped out at the different pressures was collected in the measuring tube (E), & corrected for temperature, pressure and water vapor.

These corrections are

Pressure	Vol. air collected	Pressure	Vol. air collected
460 mm	29.9 cc	130 mm	60.0 cc
360 "	39.0 "	100 "	63.2 "
260 "	48.2 "	60 "	66.4 "
160 "	57.6 "	40 "	68.0 "

The compound having been prepared in the usual manner, it was only necessary to close the connection with the gas burette, and open the stopcock leading to the pump.

In the first series of experiments dilute hydrochloric acid (was used) (1 part acid to 12 parts water) and the tabulated results are given below.

Determination No I.   Solvent 1 HCl & 12 H$_2$O

Weight Cu$_2$Cl$_2$ = 0.50580 gr.

" CO absorbed = 0.13644 "   = 109.0^{cc}

CO = 21.24 %

Pressure (height Hg in manom.)	Total vol. gas	Correction	CO given off	% Decomposition
300 mm	29.99 cc	29.9 cc	0.09 cc	0.07 %
400 "	39.40 "	39.0 "	0.40 "	0.35 %
500 "	49.20 "	48.2 "	1.00 "	0.81 %
600 "	59.80 "	51.6 "	2.20 "	1.92 %
625 "	170.16 "	59.3 "	111.86 "	78.12 %
720 "	182.01 "	68.0 "	114.01 "	100.00

For every difference of 100 mm in pressure the apparatus was allowed to stand 24 hours, to see if the height of the manometer remained constant. This was the case until 625 mm (135 mm pressure) was reached; while occasional bubbles of gas had been given off during the entire exhaustion, the bubbles now became so numerous that the liquid appeared to be boiling. The pressure varied between 135—12. mm for a week, each time the manometer was raised higher it would drop back to this pressure and stop, but at the end of this time it became constant, and as tabulated above the amount of gas collected was 111.86 cc or 2.75 cc more than was absorbed at first. The pressure being still further reduced from 135 mm to 40 mm only 2.15 cc of gas — given off, showing that, point of decomposition is at approximately 130^{mm} pressure and the decomposition is practically complete at that point.

Decomposition No II.   Solvent, 1 part H$_2$L$_2$ 12 parts H$_2$O
Weight Cu$_2$Cl$_2$ = 0.49100 gr.
" CO absorbed   0.13174 — 105.51^{cc}
CO = 21.17 %

Pressure	Total Vol. gas	Correction	CO Given off	% Decomposition
460^{mm}	30.2^{cc}	29.9^{cc}	0.3^{cc}	0.28%
360 "	39.4 "	39.0 "	0.4 "	0.37%
260 "	49.5 "	48.2 "	1.3 "	1.21%
160 "	59.6 "	57.6 "	2.0 "	1.87%
135–125 "	165.6 "	59.3 "	106.3 "	79.42%
40 "	174.9 "	68.0 "	106.9 "	100.00%

These two decompositions are given

as extremes, the first being exceptionally poor, the last being slightly better than the average. In all cases more gas was collected than the amount of carbon monoxide plus the correction to be applied, but I think one can easily understand this error when it is noted that not all rubber connections could be dispensed with, owing to the necessity of shaking the bulb during the absorption of carbon monoxide — though these connections were kept well coated with shellac. However, it hardly seems possible that this should have entirely prevented leakage through the rubber, assuming that all the stop cocks were perfectly tight, for each experiment extended over about three weeks, and for the most part a very low press. was maintained In the

The decomposition-bulb had to be kept in ice-water all this time, and of course it was a matter of considerable difficulty to maintain an exactly constant temperature during this period.

Considering then these little sources of error, I think we can overlook the variations observed, and it seems quite evident that the compound as formed in this way, - viz very dilute hydrochloric acid being used, does not decompose until the pressure is reduced to 135-125 mm, and that it is completely decomposed at 1 mm about.

<u>Behavior of the Compound as made with Water</u>

The compound obtained by the absorption of carbon monoxide in cuprous chloride in water, was next submitted

In new pressure of 0°C in quite unexpected results were obtained

Experiment No. I

CO_2 absorbed 10.0 °C.

Pressure	Total Vol. Gas	Correction	CO₂ burn Off	% Decompos
460ᵐᵐ	30.1ᶜᶜ	29.9ᶜᶜ	0.2ᶜᶜ	0.22
360 "	39.3 "	39.0 "	0.3 "	0.33
260 "	49.0 "	48.2 "	0.8 "	0.88
160 "	58.8 "	57.6 "	1.2 "	1.32
130 "	102.7 "	60.0 "	42.7 "	46.97
100 "	109.2 "	63.2 "	45.7 "	50.27
60 "	156.9 "	66.4 "	90.5 "	99.55
40 "	158.9 "	68.0 "	90.9 "	100.00

By way of explanation:— in lowering the pressure from 160ᵐᵐ to 160ᵐᵐ only 1.2ᶜᶜ of carbon monoxide were given off, but at 130ᵐᵐ pressure the volume was gas came quite very soon, and when the pressure was lowered it rose again to the point and stopped. This was continued for about five days when 42.7^{cc} of carbon monoxide had been given off, and when the pump was started the manometer rose quite rapidly and between the pressures of 130 and 60^{mm} only about 5^{cc} of gas was given off. However at about 60^{mm} pressure the evolution of gas seemed to increase and the total volume given off up to and including 60-55^{mm} was 90.5^{cc}. But 42.7^{cc} were given off at 130^{mm} and 5^{cc} between 130 & 60^{mm}, so the volume given off below 60^{mm} pressure = 90.5^{cc} - (42.7+5^{cc}) = 42.8^{cc}. The total volume given off was 90.7, so that almost exactly half was given off at 130^{mm} and the remainder at 60^{mm} pressure.

Considering the small volumes of gases worked with and the fact that the two amounts of decomposition are quite low and very close together, being 130 and 60 m.m. respectively; remembering also the low temperature employed and the great length of time required, I think, any one will agree that the figures are not altogether so rough as they appear at first sight.

Experiment No II.
   Another decomposition, water being used, gave results as follows:
   CO absorbed = 103.33 cc

Pressure	Total Vol Gas	Correction	CO_2 given off	% Decomposition
460 mm	30.2 cc	29.9 cc	0.3 cc	0.28
360 "	39.4 "	31.0 "	0.4 "	0.37
260 "	49.2 "	48.2 "	1.0 "	0.95
160 "	57.0 "	55.6 "	1.4 "	1.33
130 "	109.99 "	60.0 "	49.99 "	47.70
100 "	114.92 "	63.2 "	51.72 "	47.35
60 "	170.87 "	66.4 "	104.47 "	99.63
40 "	172.87 "	68.0 "	104.87 "	100.00

Total vol. gas collected = 104.87 cc
"   "   " absorbed   103.33 "
      Difference   1.54 "

½ Vol gas absorbed = 51.66 cc
Vol. gas given off at 130 mm = 49.99
      Difference   1.67 "

I have noted during the decomposition by induced pressure, of the compound as made with the use of water, that the cuprous chloride left in the decomposition bulb was quite dark, due to an admixture of dark particles, and there was also a very thin dark coating on the bottom of the bulb. It was not possible to tell with certainty what this material was, the quantity being too small, but it certainly contained copper, for the part attached to the bulb was carefully washed and tested; in appearance it was exactly like the coatings of metallic copper formed by its precipitation very slowly and in fine condition. I've gone as mentioned for since it was running in explanation of the formation of carbon dioxide where oxygen acts on Carbonyl nitrous chloride.

Behavior of the Compound as made with Concentrated Hydrochloric acid.

Insomuch as the conduct of the compound made with water differs from that made with dilute hydrochloric acid, it was deemed necessary to test the behavior, under reduced pressure, of the compound made with concentrated acid. The following example will suffice to show the results in a general way—

    Solvent conc. HCl.   Temperature 0°
    Vol. CO absorbed = 105.2^{cc}

Pressure	Total Vol. gas	Correction	CO_2 given off	% Decomposition
410 mm	40.7 cc	39.2 cc	1.5 cc	1.38
310 "	60.2 "	44.7 "	15.5 "	14.32
260 "	74.4 "	48.2 "	26.2 "	24.21
210 "	91.8 "	52.7 "	38.9 "	36.75
160 "	117.2 "	57.6 "	57.6 "	53.08
130 "	168.2 "	60.0 "	108.2 "	100.0

It will be seen from the above that in this instance decomposition begins at a pressure of about 410 mm and gradually increases as the pressure is lowered to 160 mm; from 160 to 130 mm the decomposition is markedly increased and at 130 mm is complete.

This agrees with those in which dilute acid is used in that decomposition is complete at 130 mm pressure, is to be from both ...

others in that way. Decomposition begins below 130 mm. The stronger the acid the greater the ease with which carbon monoxide is absorbed by cuprous chloride and the greater the ease with which this gas is again given off under the influence of reduced pressure.

Experiments in which other strengths of acid were used were not undertaken for an experiment extends through ... three weeks, the bulb containing the compound being kept packed in ice all the time, and to maintain a ... requires almost constant attention. Further, the slightest flaw in stopcocks or connections will show itself when a pressure of ... -1 mm. is main-for several weeks and of course the slightest leak means absolute loss of the experiment. On account of these experimental difficulties it was not believed that the value obtained necessarily —— to a fine degree — would vary —— time spent.

These results will later be discussion in detail and the decompositions be expressed in the form of curves.

Absorption of Carbon Monoxide by Cuprous Chloride in Pyridine.

It was desired to test the effect of increased temperature on some compound of cuprous chloride and carbon monoxide. Plainly the compound formed by the absorption of this gas by cuprous chloride in water or hydrochloric

acid is too sensitive to heat to permit
an accurate study of its decomposition
by increase of temperature.

N. Lang[1] has noticed that cuprous
chloride in pyridine has the power of
absorbing carbon monoxide. It
was tried in the hope that a compound might be obtained which would
prove more stable toward heat than
the one first studied; this com... and
was... ... the second properties

The ..... method of ... ..-
tion was the same as that described
on page 11, except that the water
in ... was replaced by pyridine.
One peculiarity was noted; no matter
how carefully the bulb was exhausted,
the instant the ... came in
contact with the cuprous chloride both

(1) Ber. 21 1584

assumed a deep green color, due probably to some product of xxx ion. The flesh color became xxx account of the xxxx between the two compounds, and the cuprous chloride exhibited a great tendency to form small hard lumps. It was necessary in every case to heat to 60 or 70°C in order to get these lumps broken up and into solution, & it was difficult to tell whether xx particles xxx not simply suspended, as the solution was too dark to see through. xxx xxx xxx xxx xxx xxx xxx possible the bulb was alternately shaken & immersed in ice-water until no more carbon monoxide could be absorbed, & in the previous instance too. The endp

the of pz=xa hay & the absorption became
extremely slow.

Various attempts were made to isolate
the compound formed, part of which was
in solution in the p=xxline, the other appearing as beautiful white plates on the
bottom of the bulb, but all these attempts
failed utterly. On the addition of ice-
water to the pyridine, more of the compound is thrown down in the form
of plates, and when it is attempted to
filter these off they ... ... in the
instant the air touches them. So much
more unstable in the air is this compound than the one made by the use
of water or hydrochloric acid, that an
attempt to dry it on drying-paper results in almost complete disappearance.
In several instances there is

added to the pyridine solution several times its volume of 50% alcohol and the mixture allowed to stand in the air for several hours. The color of the solution became very much intensely green, and gradually there were deposited beautiful crystals, similar in appearance and properties to the compound described by Lang, to which he ascribed the formula $Cu_2Cl_2 \cdot 2C_5H_5N$. The compound can be easily filtered, washed and dried and comes out in pure form. An analysis ... gave ... the following result.

0.2103 gr. material gave   0.0135 gr. CuO
Calculated for   Found
$Cu_2Cl_2 \cdot 2C_5H_5N$
Cu = 21.65%   21.61%

Drying in ... to obtain the ...

1. Ber. ..

for us, the morphology of cuprous oxide of cuprous salts in pyridine recourse must be had to the absorption method, than, of course, the only method of the determination of the proportion between the two constituents cuprous oxide and cuprous chloride.

A large number of absorptions were made and below is given an average series, though the results tend rather to the higher than to the lower percentage.

Weight Cu_2Cl_2	Weight CO absorbed	% CO
1.0005 gr.	0.17263 gr.	14.71
0.2020 "	0.035630 "	14.98
0.2055 "	0.036317 "	15.00
0.4036 "	0.012710 "	15.26
0.4028 "	0.076182 "	15.70
0.2040 "	0.037035 "	16.05

at once that the results vary quite widely, due probably to the caking of the cuprous chloride on the introduction of pyridine, and also to the tendency of the compound so formed to enclose some of the unchanged cuprous chloride, thereby making complete absorption impossible.

If the relation of cuprous chloride to carbon monoxide is a simple one, it is in all probability $2Cu_2Cl_2 \cdot 3CO$, since the percentage of carbon monoxide, as determined by the absorption method, is undoubtedly low.

The percentages required by the simpler hypothetical compounds are the following –

$Cu_2Cl_2 \cdot CO$; $CO - 12.41\%$
$2Cu_2Cl_2 \cdot 3CO$; $CO - 17.57\%$
$3Cu_2Cl_2 \cdot 5CO$; $CO - 19.10\%$
$Cu_2Cl_2 \cdot 2CO$; $CO - 22.15\%$

While this method of analysis does not make it possible to determine whether any HCl enters into the compound, yet by analogy to the compounds made with water and hydrochloric acid ($HgCl_2 \cdot 2CO \cdot 4H_2O$ —), one would be led to suspect that it does.

If the proportions $2 HgCl_2 \cdot 3CO$ is correct, and pyridene enters into the compound, it might occur to any one that it is derived from $2(HgCl_2 \cdot 2CO)$ by replacement of one CO by $C_5H_5N_2$, though there is no evidence whatever on this point.

<u>Behavior of the Compound under increased Temperatures.</u>

The method of work was the following. Into a bulb of 25 cc capacity was introduced a weighed quantity of cuprous chloride, about 0.403 gr, and the bulb closed with a two hole rubber stopper, through one opening was fixed a dropping funnel and through the other a bent glass tube. The bulb was then alternately exhausted and filled with nitrogen until it was completely filled with this gas; 5 cc of pyridine was accurately measured and introduced through the dropping-funnel, the bulb was connected by means of a capillary tube with a gas burette, and the bulb immersed to a fixed mark in a water bath, the temperature of which was accurately regulated.

The temperature of the bath was gradually increased, readings being made at the desired points. Before making the final readings of the volume of gas in the burette, and the immersing of the tip in the bulb the temperature we kept at the same as for an hour, in order that the volume may be absolutely constant. In this way a series of corrections was obtained, which was applied in the actual determinations. The corrections used, the average of several series, were the following:

Temp of bath	Vol. gas collected	Temp of bath	Vol. gas collected
0° C	0.000 cc	60°	8.572
10°	1.869 "	70°	10.863
20°	2.849 "	80°	13.578 "
30°	3.917 "	90°	16.737 "
40°	5.164 "	100°	21.401
50°	6.167 "		

As will be noticed the volume of gas collected above 50 or 60° increases quite rapidly, especially is this true as 90°–100° is the boiling point of the liquid is being approached. Also, the pyridine attacks the rubber stopper and connections at this temperature, and there may be a small error on this account.

In the actual experiments it was always impossible to get off as much gas as was absorbed; this was influenced no doubt, by the difference in the vapor tension of pyridine containing cuprous chloride and pyridine containing the carbonyl cuprous chloride pyridine compound. Indeed the action between the pyridine and rubber connections at this rather elevated temperature, or the pyridine and cuprous chloride or

This letter and carbon monoxide, or in fact all of them combined, may have affected slightly the readings obtained.

It is worthy of notice that after each decomposition the bulb contained a coating of what seemed to be metallic copper, and which ran the test for copper. The coating was more abundant than that noticed in the decomposition of the cartonyl cuprous chloride by induced pressure. It is conceivable that the following reaction is taking place —

$Cu_2Cl_2 + CO \rightarrow 2Cu + COCl_2$; and then
$COCl_2 + H_2O \rightarrow 2HCl + CO_2$.

If this was really the case, of course the carbon dioxide formed was dissolved in the water in the pev-urette, making the readings slightly low.

In all these experiments the amoun-

was present in amount corrected in
the description were made nearly as
in the blank experiments.

Decomposition, No. I.
Weight $Reg\,O_2^{\frac{1}{2}}$ = 0.4036 gr.
" CO_2 absorbed = 0.0727 = 38.144 %
CO_2 = 15.26 %

Temp.	Total vol. gas	Correction	CO_2 vol. %	% decomposition
10°	2.717 cc	1.869 cc	0.848 cc	1.45
20°	13.858 "	2.847 "	11.007 "	18.73
30°	22.283 "	3.717 "	18.366 "	31.58
40°	31.188 "	5.164 "	26.024 "	44.75
50°	40.012 "	6.767 "	33.245 "	57.11
60°	47.565 "	8.512 "	38.175 "	67.02
70°	54.511 "	10.803 "	43.708 "	75.17
80°	61.207 "	13.518 "	47.621 "	81.16
90°	68.824 "	16.737 "	52.085 "	87.57
100°	76.027 "	21.401 "	54.628 "	93.75

Decomp runs No II.
Weight $Ag_2C_2O_4$ = 0.4042 g
" CO_2 evolved = 0.6162 = 56.1376$^?$

$CO = 14.74\%$

Temp	Total Vol gas	Correction	CO giving off	% decomposing
10°	4.016 cc	1.869 cc	2.147 cc	3.82
20°	11.847"	2.847"	9.000"	16.02
30°	23.138"	3.917"	19.221"	34.22
40°	28.158"	5.164"	22.994"	40.94
50°	39.015"	6.767"	32.248"	57.42
60°	41.934"	8.592"	37.342"	70.05
70°	53.951"	10.863"	43.088"	76.72
80°	61.831"	13.518"	48.253"	85.92
90°	68.042"	16.739"	51.303"	91.35
100°	74.575"	21.401"	53.174"	94.71

Decomposition No III.

Weight $Au_2C_2 = 0.40280$ g.

" CO absorbed $= 0.07618$" — 60.7216^c.

$CO = 15.90\%$

Temp	Total vol. gas	Correction	CO given off	% Decomposition
20°	12.249^{cc}	2.849^{cc}	7.400^{cc}	15.42
40°	32.370"	5.164"	27.206"	44.65
60°	47.892"	8.592"	39.300"	64.51
80°	64.043"	13.578"	50.465"	82.83
100°	77.526"	21.401"	56.125"	92.12

Curves representing these decomposition will be found on page 134.

Action of Oxygen on Dicarbonyl Cuprous Chloride.

Experiments were next undertaken to determine whether or not the carbon monoxide at the moment of its liberation from the compound, differs in real velocity from it is ordinarily known, and also to study the ease with which it is given off when the compound is submitted to the influence of other gases.

The method of work was the following: about 500cc of carbon monoxide was absorbed by cuprous chloride just as in the preparation of dicarbonyl cuprous chloride, except that the tube through which carbon monoxide entered, was long enough to reach to the bottom of the bulb when pushed down.

The absorption complete, the bulb was introduced into the following system of wash bottles.

A	B	C	D	E	F
KOH	KOH	BaOH		H₂(OH)₂	KOH

As is shown, the oxygen was passed through two gas wash bottles containing strong caustic potash solution, through concentrated baryta water, and into the flask containing dicarbonyl cuprous chloride, the inlet tube reaching to the bottom of the flask. The gas on issuing from the flask passed through concentrated baryta water, which was protected from the air by a bottle containing caustic potash.

While the flask containing the dicarbonyl cuprous chloride was still immersed in ice-water, oxygen was admitted and almost instantly began to turn grey and bubbles of carbon monoxide were given. After the passage of oxygen for about ten minutes a decided coating invariably appeared on the inlet tube of the baryta bottle E

and at the end of half an hour a
recorded precipitate had appeared, enough
to make the baryta solution turbid when
shaken. The baryta water in the bottle (C)
before the decomposition flask remained
as clear as distilled water. This experiment
was repeated a number of times, always
with the same result.

Attempts were made to determine the
amount of barium carbonate, by weighing the sulphate and also by titration, us-
ing phenolphthalein and methyl orange
as indicators. The amount of barium
carbonate, as compared with the barium
hydrate, was so small, that an accurate determination was impossible, but
it would seem that the amount of barium carbonate did not represent an
oxidation of more than 1/10 of the carbon monoxide used.

from inquiry must be made whether the carbon dioxide came from any other source than the oxidation of carbon monoxide. It could not have been contained in the carbon monoxide for this was carefully tested in every experiment. However, only direct experiment could tell whether the cuprous chloride contained any thing which might give carbon dioxide under these conditions; another possibility was that by oxidizing the cuprous chloride the oxygen had become possessed of any activity enabling it to attack the rubber stopper and connections.

Thinking that carbon monoxide might be alternately absorbed and given off when a mixture of this gas and oxygen is allowed to pass through moist cuprous chloride at 0°C, the following method of work was adopted. From 1.62 grams of cuprous chloride were placed in the flask before described, covered with water free from carbon dioxide, the flask immersed in ice-water and placed in the series of wash-bottles as described. Pure oxygen was passed for 1½ to 2 hours, the cuprous chloride gradually becoming grey, though not as rapidly as when dicarbonyl cuprous chloride was used; the test baryta solution (E) remained absolutely clear. They without disturbing the apparatus, carbon monoxide was mixed with the oxygen in equal proportions, and when this mixture had been passed for fifteen minutes a cooling began to appear in the inlet tube of the in ... [illegible] ... At the end of two hours a perceptible precipitate had appeared.

As one might expect, the precipitate formed in this way was not as large in proportion to the carbon monoxide used, as when the carbon monoxide had been previously absorbed by the cuprous chloride; even after several more hours the precipitate did not increase very markedly, due, it would seem, to the protection of the cuprous chloride by the oxidation products formed. These experiments were repeated over and over, and always with the same result.

I might mention here that the passage of a mixture of air and carbon monoxide over an antimony or chloride [illegible] was not brought [illegible] formation of any cuprous dioxide.

Action of Chlorine on Dicarbonyl Cuprous Chloride

If carbon monoxide at the moment of liberation from dicarbonyl cuprous chloride is in a state of activity greater than that exhibited under ordinary conditions, we certainly might expect this activity to assert itself more markedly under the influence of chlorine, with which it unites with some ease.

Accordingly, experiments were undertaken in the hope of throwing some light on this point. Inasmuch as dicarbonyl cuprous chloride contains water of crystallization, phosgene, carbonyl chloride is formed, water being given off as carbon dioxide and hydrochloric cert. The determination of the amount of carbon monoxide changed under these circumstances is beset with many difficulties as a moments reflection will show, for in the absorption of carbon monoxide the vessel must be shaken, and at no time must the temperature be raised or the pressure reduced, neither must the compound be exposed to the air.

The method finally adopted, and for which great accuracy cannot be claimed, was the following:

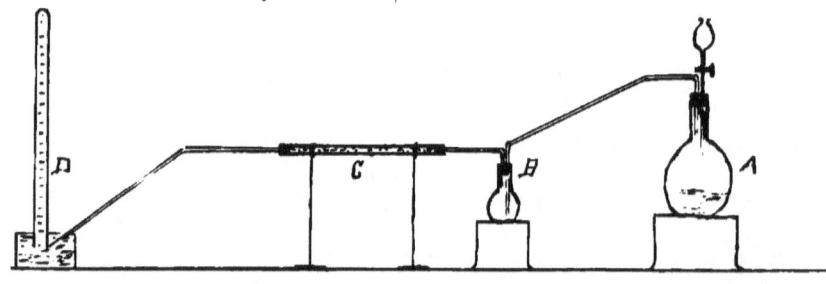

The dicarbonyl cuprous chloride was

prepared in the usual manner in a bulb of known size (about 25cc capacity) and the amount of carbon monoxide absorbed was accurately measured. The absorption-bulb was provided with an inlet tube extending to the bottom of the vessel, and an outlet tube reaching just through the stopper.

After the separation of the compound the inlet tube was connected with a chlorine generator (A) which contained as pure chlorine as it was possible to obtain; at the same instant the outlet tube was opened, this having been connected with the larger tube (C) of about 30 cm length, and containing finely divided antimony. The antimony tube was in turn connected with a gas measuring tube (D) containing strong caustic potash.

The moment the chlorine touched the crystals of dehydrous cuprous chloride they were colored grey and bubbles of carbon monoxide were seen rising from them. The passage of chlorine was continued until all the cuprous chloride had been oxidized and had gone into solution. The excess of chlorine was taken up by the antimony, and any carbon dioxide formed, or any hydrochloric acid passing over, was absorbed by the caustic potash solution; the unchanged carbon monoxide passed on and was collected.

At the end of the operation the vessel (H) containing originally the dehydrous cuprous chloride, in ... in which was dissolved, was filled with pure water so as to be sure that all ... very moment ...

is being known out. Knowing the amount of carbon monoxide absorbed and also the capacity of the absorption bulb and hence the volume of pure carbon monoxide contained in it after the absorption, it was only necessary to subtract the volume of gas collected from the sum of these two to determine the amount of carbon monoxide that had been changed.

Blank experiments were made to see if any hydrogen was given off from the antimony by the action of the hydrochloric acid which might have been swept along by the current of gas. There was found no reason to suppose that hydrogen was evolved. It was determined that pure antimony does not act as the source of gas in any known antimony chloride and antimony monoxide when a mixture of the two is passed over it.

Although the antimony tube was filled as completely as possible and all connections were made with capillary tubing, in spite also of the great care employed in having all parts of the apparatus at the same temperature before and after each decomposition, yet it was impossible to obtain very satisfactory results.

The rate of flow of the chlorine seemed to have some effect on the change of antimony monoxide, being greater the slower the current, the same effect was produced by lowering the temperature, thus retarding the decomposition of the compound.

The greatest difficulty was experienced in obtaining perfectly pure chlorine, and a very slight trace of an inactive gas would accumulate and cause a considerable error, for the chlorine had to be passed for about an hour before complete decomposition was insured.

The simplest and best method for preparing the chlorine seemed to be by the action of concentrated hydrochloric acid on sodium or potassium bichromate, but in preparing it in this way and allowing the action to proceed vigorously for two hours I found the chlorine to contain a trace of some gas which was neither taken up by the antimony nor absorbed by the caustic potash. ~~However~~ The amount

of the ... ... to ...
and two cubic centimeters;
a correction of 1.5 cc in the determinations was applied.
Below I give the average series
of determinations.

CO absorbed	CO collected	CO altered	% Change
63.2 cc	62.4 cc	0.8 cc	1.26
54.1 "	52.0 "	2.1 "	3.88
67.0 "	68.4 "	0.6 "	0.86
78.6 "	76.6 "	2.0 "	2.54
77.2 "	73.1 "	4.1 "	5.3

From these results it would
seem that there is a slight reaction between the two gases, but if
the carbon monoxide possesses an unusual activity I am sure we
should expect a much greater change in

The above figures are indicate..

Action of Hydrogen on Dicarbonyl Cuprous Chloride

Inasmuch as oxygen has the power to decompose this compound even at or below 0°C., it seemed advisable to try the action of hydrogen, thinking that if the hydrogen did decompose it such a reducing product as ~~methyl~~ formyl aldehyde-mig. be formed.

The compound was prepared in the usual manner and the bulb containing it, still immersed in ice water, was introduced into a system of gas wash-bottles similar to that described on page 67. The hydrogen was prepared by the action of hydrochloric acid on pure zinc and then washed with caustic potash, no acetate on potassium permanganate. The hydrogen was bubbled through the compound (at 0°C) for eight hours and at the end of this time the crystals had in no way changed their appearance.

The water contained in the wash bottles through which the gas passed on leaving the wash bottle containing the compound, was examined carefully; it had neither taste nor odor, was neutral in reaction and did not reduce Tollens' solution in the cold.

The bulb containing the decarbonyl cuprous chloride was finally allowed to assume the room temperature, when the carbon monoxide was slowly given off, and later the bulb was heated to 60°; but no change in the carbon monoxide could be detected.

Action of Nitrogen.

These experiments were conducted exactly as in the case of hydrogen. At 0° the compound was not affected by the nitrogen, and when the bulb was warmed so that the carbon monoxide was given off, no change whatever could be detected in the latter.

Action of Acetylene on Carbon Monoxide

When acetylene is introduced into water gas for the purpose of enriching it, a proportionately large amount of the former has to be introduced before the flame shows any [...] mostly whatever.

In decarbonyl cuprous chloride we have a union of two unsaturated compounds, each of them having two unsaturated valences; since acetylene is also an unsaturated compound and to exactly the same extent as the two compounds just mentioned, it seemed possible that carbon monoxide and acetylene might unite under certain conditions and in this way the difficulty experienced in enriching water gas with acetylene be explained.

Suffice it to say that all the results obtained were distinctly negative. Measured volumes of carbon monoxide and acetylene were mixture over mercury and allowed to stand for a week at the ordinary temperature exhibits no change in volume. The mixture when passed through an ammoniacal solution of cuprous chloride precipitates the characteristic copper carbide and the entire amount of gas can be absorbed by cuprous chloride.

A similar mixture passed through a heated tube, suffers no change except that due to the action of heat on acetylene.

The passage of acetylene into dicarbonyl cuprous chloride, or a mixture of acetylene and carbon monoxide into cuprous chloride, or treatment of a mixture of copper carbide in conbonyl cuprous chloride with hydrochloric acid, all fail to effect a union of acetylene and carbon monoxide.

Some experiments were conducted to determine just what proportion of acetylene is necessary to render a carbon monoxide flame luminous. Mixtures of the two were made and burned from a platinum tip, the proportion of acetylene being gradually increased until the flame became very slightly luminous.

Experiments were also conducted with mixtures of acetylene and hydrogen, and acetylene and water-gas rendered non-luminous by being passed through fuming sulf. a (see "fuel gas" below).

The results are the

C_2H_2 and CO ; 11% C_2H_2 required
C_2H_2 and H ; 1% C_2H_2 required
C_2H_2 and "fuel-gas"; $2.-2.5\%$ C_2H_2 required

The case of acetylene and carbon monoxide is not an unusual one; instead of the acetylene enriching the carbon monoxide, this latter renders the acetylene flame non luminous, just as hydrogen does any ordinary gas flame. Of course this is not an explanation, but it is in accordance with facts previously noted and so excellently discussed by Heumann as far back as 1876.

The nature of the combustion of acetylene and the rôle it plays in luminous flames, has recently been discussed quite fully by Prof. Lewes[2], so that

1) Annalen (Liebigs) 181, 129; 182, 1; 183, 102 + 184, 201.
2) Proc. Roy. Soc. 57, 45.

Discussion of Results.

Action of Oxygen on Dicarbonyl Cuprous Chloride.

There can be no doubt that when oxygen is passed into this compound even at 0°C, it undergoes decomposition and there is formed simultaneously a small amount of carbon dioxide. This carbon dioxide cannot be due to an impurity in the cuprous chloride, for oxygen passed through the same cuprous chloride, either at 0° or 60°, gives no carbon dioxide whatever. Further it cannot be due to a peculiar activity of the O_2;

brought about by the cuprous chloride through its oxidation thus causing a splitting of the molecular into atomic or "active" oxygen. For when a mixture of carbon monoxide and oxygen is passed through ferrous sulphate or chloride these compounds are oxidized yet none of the carbon monoxide is converted into carbon dioxide.

Plainly there the oxidation of carbon monoxide by the action of oxygen in dicarbonyl cuprous chloride must be due either to the activity of the carbon monoxide or else it must in some way be brought about through the agency of the cuprous chloride.

While we have some evidence pointing to the fact that both elements and compounds are unusually ac-live at the moment of their iora-tion, yet it is far from conclusive that this activity is a property of the element or compound per se. Certain heat phenomena are involved in all chemical reactions and in some instances at least electrical phenomena, so that an explanation of the real or apparent activity of the "nascent state" must be at present very much involved.

In the experiment under discussion one point should be noted especially; if the formation of carbon dioxide is a result of the oxidation of the carbon monoxide on account of its activity, then it is very surprising that chlorine, which unites with carbon monoxide with some ease, does not do so to a much greater extent under these conditions and yet it does not.

Plainly then, to explain this oxidation we have recourse to only one line of investigation and we may ask in all seriousness what part is played by the cuprous chloride, or the products of its oxidation which are necessarily present.

Drehschmidt has shown[1] that when an ammoniacal solution of cuprous chloride is used for the absorption of carbon monoxide, a coating of metallic copper is formed on the absorption pipette, due, as he thinks, to a reduction of the cuprous chloride by the carbon monoxide.

Further Winkler shows[2] that caustic potash causes the precipitation of metallic copper from a solution of cuprous chloride containing carbon monoxide; carbon dioxide is formed at the same time. Winkler also points out[1] that by the action of carbon monoxide on a solution of cuprous chloride and palladium chloride there is a precipitation of palladium and formation of carbon dioxide, thus

$PdCl_2 + CO + H_2O = Pd + CO_2 + 2HCl$

He also shows that carbon monoxide acts very slowly on a solution of palladium chloride, the cuprous chloride facilitating the action very markedly.

I wish to recall here that in the decomposition by reduced pressure of dicarbonyl cuprous chloride suspended in water, there was always ob-

(1) Ibid 28, 214

secured a slight black coating on the flask, having the appearance of metallic copper and giving the copper test. A still heavier coating was obtained in the decomposition, by increased temperature, of the contents or by the absorption of carbon monoxide by cuprous oxide in pyridine, though in this case it is possible that the pyridine exerts a reducting influence as well as the carbon monoxide.

I have frequently noticed, as no doubt have others, that when strong hydrochloric acid is added to cuprous chloride, the latter turns dark, and when this is used to absorb carbon monoxide, it becomes darker until this optical effect is partially overcome by the bright crystals of dicarbonyl cuprous chloride formed.

Phillips has shown[11] that by the action of carbon monoxide on solutions of the chloride of palladium, iridium, platinum and gold at the ordinary temperature, and on a solution of rhodium chloride at 100°C, the metals are precipitated and carbon dioxide formed.

Berthelot[a] has shown, and it has been confirmed by Phillips[13], that when carbon monoxide is passed into an ammoniacal solution of silver nitrate at the ordinary temperature, metallic silver is deposited.

Looked at in the light of the periodic law these facts are of unusual interest. Copper belongs to the same sub group as do silver and gold; now the peculiar transition group just before copper (63) and acting as a kind of connecting link between manganese (55) of the 4th series and copper of the 5th series, comprises iron (56), cobalt (59) and nickel (59), of which both iron and nickel form compounds with carbon monoxide. Again the transition group ruthenium (104), rhodium (104) and palladium (106) connects the 6th series with the 7th, the first member of the latter being silver (108); notice that carbon monoxide quickly reduces potassium ruthenate to metallic ruthenium, also reduces rhodium chloride at 100° and palladium chloride at the ordinary temperature, it also reduces an am moniacal silver nitrate solution

Passing downward again , find the Transition group osmium 191, iridium 193', and platinum 195, connecting the 10th and 11th series, the first member of the latter being gold 197. Carbon monoxide quickly reduces osmic acid and also reduces the chlorides of iridium, platinum and gold to the metals, carbon dioxide being formed.

Of all the metals comprising the transition groups, Ir, Co, Ni, Ru, Rh, Pd, Os, Ir, Pt, some of their compounds, especially the chlorides, are reduced by carbon monoxide, carbon dioxide being formed, with the exception of cobalt and nickel, and nickel has the distinction of uniting with carbon monoxide. Of the sub group, Cu, Ag, Au, the nitrate of silver is easily reduced as is the chloride of gold, the chloride of copper being the only exception. Notice further that this peculiar behavior under the influence of carbon monoxide at ordinary temperatures, is confined almost, if not entirely, to these metals.

Of course, we cannot reason conclusively from relations brought out by the periodic law, yet this law does lend quite a degree of probability, in conjunction with the experiments of Dr. Schmidt and Winkler, as well as the phenomena described by me. Especially is the probability increased when we are forced to replace this formation of carbon dioxide by the activity of the carbon monoxide or else by the influence exerted by the cuprous chloride. I think any one would be inclined to represent the reaction as Winkler has shown in the case of palladium chloride: $PdCl_2 + CO + H_2O = Pd + CO_2 + 2HCl$ and correspondingly $Cu_2Cl_2 + CO + H_2O = 2Cu + CO_2 + 2HCl$

The Constitution and Chemical Nature of Dicarbonyl Cuprous Chloride.

As regards the constitution of this compound I have no suggestion to offer. Formulae might be written to satisfy all the bonds of the elements involved, but there is no experimental basis for such a procedure and it would add nothing to our knowledge. It seems fairly certain that its composition is represented by the formula $Cu_2Cl_2 \cdot 2CO \cdot 4H_2O$, remembering that both cuprous chloride and carbon monoxide are unsaturated compounds, it seems only natural that they should be able to unite and mutually saturate each other, but how the parts are joined together is a far more difficult problem.

What is still more important under the circumstances, and as regards which I think there is at least some weighty evidence, is a consideration as to whether the so-called compound is in reality a true chemical individual, an "atomic" compound. So far as appearances are concerned, — the crystals be-

seemingly homogeneous and charac-
terized by a constant crystal form
and lustre - damn sure any one would
be inclined to consider it a chemical
individual. The compound is quite
stable at low temperatures, in the absence of active substances, but if the
temperature be elevated or the pressure
greatly reduced it undergoes decomposition.

While the term "molecular compound"
would most assuredly have been applied to this compound some years ago —
a term then indefinite enough but
now so utterly indefinite as to mean
almost nothing unless it be a term
to designate our ignorance of the cause
of a certain instability observed in
some compounds - Yet even at the

present time there is at least some tendency to set apart certain compounds as differing from the true atomic compounds. I cannot but feel, judging from the utterances of both books and journals, that some chemists will be inclined to place this compound in that category.

It may be well to recall just how the term "molecular compound" came into use, the conception that is responsible for its existence, the aid it has rendered to the advancement of our science and finally to consider the ground on which it now stands, that we may see whether it has any true meaning and value at present.

The conception of "molecular" as opposed to "atomic" compounds was the outcome of the conception of the saturation capacity of elements and the necessary result of the idea of constant valency. During the development of the type and radical theories in the first part of this century, it was repeatedly pointed out that in all cases of substitution a definite number of atoms of the one element is replaced by a definite number of atoms of the second.

Dumas in 1834 had pointed out that in the substitution of hydrogen by chlorine, one **atom** of the former is replaced by one atom of the latter, while two atoms of chlorine are required to replace one atom of oxygen.

Indeed the four fundamental types themselves were steps taken in

seriously in the same direction, the
valency rising from one to four in the
order in which they are given

$Cl\{H$; $O\{^H_H$; $N\{^H_{HH}$; $C\{^{HH}_{HH}$

Further the different replacement values of metals were brought out by the
classic work of Graham on the phos-
phoric and arsenic acid, and also
by Liebig in his work a little later on
the polybasic organic acids. The full
signif cance of these facts was not recognized, however at that time, and we may
say that our conception of valency dates
from the appearance of Frankland's
paper[1] in 1852, "On a new Series of Organic Compounds Containing Metals."
He showed that while Sn has the power
to form two oxides, SnO and SnO_2, yet

[1] Phil. Trans. 142, 4.

diaethylium, $Zn(CH_3)_2$ incapable of forming only one oxide, $Zn(CH_3)_2O$, the zinc here appearing to be saturated, and this compound may be looked upon as $ZnSO_4$, one oxygen of which has been replaced by $(CH_3)_2$.; similar results were obtained with other organo-metallic derivatives. From a comparison of the oxides, chlorides etc of phosphorus, arsenic, antimony and nitrogen, Frankland showed the tendency of these elements to unite with other elements in compounds of fixed or ─ atoms of the uniting elements. Thus is   that Frankland assumed a saturation capacity of the elements, the limit in the cases cited by him being exceed by the number 5. However he did not suppose that the elements have a constant valency; his idea was that although valence is variable as we're seen in the compounds $SnCl_2$ and $SnCl_4$, yet in every case there is a limit, and while in the case of PCl_3 the saturation capacity has not been reached, yet there is a limit, and this limit is reached, as he thought, in the compound PCl_5.

It is but fair to state that the work of Odling, a little later, lent a great deal of support to these ideas and it was he who first made use of the term "replaceable value." He claimed that both iron and tin have two replaceable values and introduced the symbols, Fe'' and Fe''', Sn'' and Sn'''', still in use.

These ideas seemed well established during the fifth decade, such elements as Friedmeyer, Kolbe, Wurz, Williamson and Gerhardt accepted them, and ammonium chloride was looked upon as a saturated compound, a compound in which all five of the valencies of nitrogen are expressed. Ammonia was considered just as truly a chemical compound, except in this instance only three of the five uniting affinities come into play, the compound thus being unsatd.

In the year 1857 Kekulé[1] introduced the conception of mixed types, a modification of the condensed type of Williamson and in 1858 he established the tetratomicity of carbon[2], introduced the marsh gas type and established his doctrine of "Binding der Atome".

[1] Annalen der Chemie, 104, 129
[2] Ibid. 106, 121

Kekulé's ideas underwent a gradual growth as can be seen from his papers in Liebig's Annalen during 1857, '58 and '59 and culminates in his famous "Lehrbuch der Organischen Chemie" which began to appear in 1859. One can well see how that all the work of Kekulé, from that on the sulphonic acids to that which led him to his conception of the linking of atoms, brought him, by a peculiar combination of facts, to believe that the valency of elements is constant, and indeed "he declares" it to be as unalterable as the atomic weights.

This led to quite an animated discussion during the years 1863, '64, between Kekulé and Wurtz, Naquet and others. Kekule looked upon nitrogen and phosphorus as only trivalent, sulphur and oxygen bivalent etc. and in order to explain the compounds NH_4Cl, PCl_5 etc. he was forced to assume a fundamental difference between NH_3 and PCl_3, and NH_4Cl and PCl_5; The former he termed "atomic" the latter "molecular" compounds. It was known that the vapor densities of these latter compounds were of an inexplicable kind and the criterion by which he judged a compound to be atomic or molecular was its ability or inability to be converted into the gaseous state without decomposition. He claimed that in the atomic compounds, the component parts are held together by the attractive affinity of the individual atoms; this might be termed the interatomic force. The molecular compounds, on the other hand, were supposed to be made up of two or more individual molecules, which were held together by an intra-molecular rather than an interatomic force, which, however, is sufficiently strong to preserve the union under the ordinary conditions.

It can be seen at once how artificial and meaningless such a division of compounds must have proven to be, and it is not unnatural that great opposition was encountered. That this was the case can be seen from the articles of Kolbe[1] and Naquet[2]. One sentence from the paper of Kolbe, just referred to, is worthy of repetition since it represents the case so admirably; speaking of molecular compounds he adds, "It is, to be sure, a word and so sounds very learned, however it is only a word, as that of which the poet says: 'Wo die Begriffe fehlen, da stellt ein Wort zu rechter Zeit sich ein.'"

Yet in spite of all the opposition that was stirred up against this artificial separation of compounds into the "atomic" and "molecular" varieties, in spite of the convincing proofs that such a procedure is unwarranted, just the same there has crept into the minds of some chemists a belief that there is an internal difference in structure in some compounds, just as Kekulé first suggested — a difference expressing itself in ease of decomposition. True, great ease of decomposition, e.g. under the influence of heat, must be an evidence of comparative instability in the compound where heat changes are involved, but does this justify one in calling such a compound "molecular"? Yet this is exactly what is not infrequently done, in proof of which I have only to refer to the often quoted book of Neumann "Ueber Molekülverbindungen nach festen Verhältnissen" Heidelberg 1872, in which he makes exactly the same distinction as did Kekulé ten years previously, or to come to the immediate present I cite the "Lehrbuch d. Chemie" of Meyer and Jacobson[1] in which the

[1] Vol. I pp 58, 59

terms the compound formed by methyl ether and hydrochloric acid "molecular," being made up $(CH_3)_2 O \cdot HCl$. And just as until very recently the so-called "double halides" were considered molecular compounds and even now the salts containing water of crystallization are quite frequently placed in this class.

The survival of this conception was due probably, in the first instance, to the fact that it was put forward by Kekulé and incorporated in his "Lehrbuch", and naturally the great brilliancy of his   K, as regards the structure and relations of organic compounds, gave a great and lasting impetus to his conception of molecular compounds. Then too, the book of Naumann just referred to, did something to sustain the idea, but more potent than all else was that peculiarity of human nature which Kolbe evidently recognized - a strong tendency to substitute a term for lack of knowledge and thus lull the conscience to rest and finally to be so deceived as to consider the term knowledge.

The state of chemical knowledge happened to be peculiarly advantageous for the ideas put forward by Kekulé. It so happened that not a single compound was known, in which a valency seems to exist greater than that granted by Kekulé, but also showed an abnormal vapor density, and many of the proofs we now have were then lacking.

Two of the cases especially cited by

Kekulé were the abnormal vapor densities of ammonium chloride and phosphorous pentachloride. The irregularity observed in the densities of these compounds was assumed to be due to their decomposition when converted into vapor.

$NH_4Cl \rightarrow NH_3 + HCl$, and $PCl_5 \rightarrow PCl_3 + Cl_2$

This assumption received absolute proof when in 1862 Pebal[1] succeeded in separating the vapor of ammonium chloride into ammonia and hydrochloric acid. This result was obtained by utilizing the difference in the rate of diffusion, through an asbestos diaphragm in hydrogen, of the lighter ammonia as compared with the heavier hydrochloric acid.

Similarly, and without the use of a porous diaphragm, which nature

[1] Annalen der Chemie 123, 199

has been objected to. Wanklyn and Robinson" succeeded in separating phosphorus pentachloride into phosphorus trichloride and chlorine, at a temperature of 300°C. The diffusion taking place in an atmosphere of carbon dioxide. It is interesting to note that the same experimenters succeeded in separating sulphuric acid into sulphur trioxide and water by means of the same method, the temperature being in this case 445°; run by sulphuric acid is a true chemical compound.

Undoubtedly this work did lend great aid to the conception that in ammonium chloride and phosphorus pentachloride, the affinities of the nitrogen are stronger than the other two, in the former instance and that in the latter case the same holds true for the affinities of phosphorus.

That this does not constitute such a difference of internal structure as claimed by Kekulé, was very ingeniously shown by Meyer and Preis" by the use of a substituted ammonium chloride. One of the compounds they made use of was dimethyldiethylammonium chloride; plainly if there is any difference in the relations which these constituents have to the nitrogen atom, then the compound as made in this way

$$N(CH_3)_2 C_2H_5 + C_2H_5 Cl = N(CH_3)_2 (C_2H_5)_2 Cl$$

should differ from the one made thus

$$N(C_2H_5)_2 CH_3 + CH_3 Cl = N(C_2H_5)_2 (CH_3)_2 Cl.$$

However, not the slightest differences

was found & the same result was reached in the case of any substituted ammonium chloride then. There seems to be no doubt that in these compounds nitrogen is pentavalent and yet their decomposition is exactly analogous to that of ammonium chloride:

$NH_4Cl \rightarrow NH_3 + HCl$
and $NR_4Cl \rightarrow NR_3 + RCl$

It is but fair to state that the question has been raised by Losser, with some reason, whether the methods employed by Meyer and x... are sufficiently refined to reveal such slight differences in constitution as might be caused by a difference in the affinities of the two atoms of a compound as contrasted with the other three.

Utilizing the same principle as that employed by Meyer and Lrcco, Collie[?] has shown that in the substituted phosphonium compounds all the organic radicals bear the same relation to the phosphorus atom. For example, he found that di-methyldiethyl phosphonium iodide as prepared in these two ways are identical - and so cannot be "molecular."

$$P(C_2H_5)(CH_3)_2 + CH_3 = P(C_2H_5)(CH_3)_2 \cdot CH_3 I$$
$$and \; P(CH_3)_3 + C_2H_5 I = P(CH_3)_3 \cdot C_2H_5 I$$

In 1870 Thorpe[12] prepared phosphorus pentafluoride which is a gas, and its density and hence molecular weight agrees with that required by the formula PF_5. This work has been confirmed by Moissan[11], the mean of his vapor density determinations being 4.49, that required by PF_5 being 4.40

Further Baker claims that the vapor density of absolutely dry ammonium chloride is not abnormal at a temperature of 350°; the mean of six determinations being 27.8 while that required by NH_4Cl is 26.75. If however, the least trace of moisture be present, the value drops at once to 13.2 – 13.9.

I do not see how these results can form any basis for the distinction made by Kékulé between molecular and atomic compounds. However, some may feel that a compound that is decomposable by a reduction of pressure must in some way differ from a true chemical compound; the marked effects of heat changes are so apparent as in

Jour chem Soc. 65 615

to be mistaken that the effects produced by a serie change [...] need to almost [overlooked]; on this account I propose to give a few examples, since if the compounds at least, or say in truly atomic compounds as say we know.

Effect of Pressure Changes on the Molecular Weight of Compounds

It has been observed by Naumann[1] that not only does the density of acetic acid vary with the temperature, but also with the pressure, the temperature remaining constant. One series of determinations will suffice to show this

Ac: Acid		Temperature 100°	
Pressure	Density	Pressure	Density
393 ᵐᵐ	3.44	258 ᵐᵐ	3.17
342 "	3.37	232 "	3.12

(1) Annalen der Chemie 155, 325

Pressure	Density	Pressure	Density
186 mm	3.06	130 mm	2.94
168 "	3.01	92 "	2.76
156 "	2.98	78 "	2.66

Theor. heat for $C_2H_4O_2 = 2.073$
" " $C_4H_8O_4$ 4.150

That these results are not confined to this temperature is shown by the fact that similar results were obtained at temperatures ranging from 78° to 185°. This work is confirmed by that of Ramsay and Young, results of the same order being obtained through a much wider range of pressures.

Quite an interesting example is that of hydrofluoric acid. The molecular weight as determined by Gore corresponds to the formula HF. In 1881 Mallet determined the density of the vapor of this acid at 30° by

weighing known volumes of its vapor, d
obtained 39.32 for its molecular weight,
ag[r]eeing almost exactly with that required
... H_2F_2.

By a still more extensive piece of work Thorpe
and Hambly" show that u[nder] the ordinary temperature the compound is to
be expressed by the formula HF, by lowering the temperature the molecule becomes
more and more complex until it ap-
proaches H_3F_3, and this change is
gradual, there being no breaks in the
vapor density determinations. To reverse
this, an elevation of temperature exhibits
a disruptive influence and they found
that a decrease of pressure exerted a similar
influence as can be seen from the
followi[ng] series of determinations

Temperature	Initial Pressure	Vapor Density (H)	Comp. Molec Wt.
32.0°C	743ᵐᵐ	19.87	39.74
32.2°	686 "	17.87	35.78
3.8°	655 "	16.99	33.98
32.0°	603 "	15.41	30.82
32.5°	545 "	13.89	27.78
32.3°	498 "	13.27	26.54
31.9°	354 "	11.50	23.00
32.3°	353 "	11.39	22.78

The case of nitrogen peroxide is of the same order. Nau... ann has shown that for a given temperature the dissociation of N_2O_4 into $2NO_2$ is greater the less the pressure, as these two cases show

Temp. 18.0°C and 279ᵐᵐ press'r   "   a joy = 1.5 :
 " 8.5° and 136"   "   "   = 29.8
Same Temp. 20.0° and 301'   "   "   = 17.8
  " 20.8° and 153"   "   "   = 29.3

Ber. 11, 2045

One of the most peculiar cases "$\ldots$" is
that first noticed by Bohr". When a given
mass of oxygen is enclosed in a tube and
the pressure reduced to about 0.1mm, a state
of instability seems to be reached, the volume
of gas tends to increase as though some
of the molecules had been split into
smaller particles. The same phenomenon
has also been noted by Baly and Ramsay"
and also by R. Threlfall", but no satisfactory explanation has been reached.

The Disruptive Influence of Reduced Pressure on Chemical Combinations.

One of the most striking instances of this
kind is the effect produced by reduced
pressure on the dissociation of hydriodic
acid as is shown by the work of ———
The main point established is that

with constant temperature the less the
pressure the smaller the amount of
hydrogen and iodine in combination
for example — the values obtained
at 440°.

Pressure	of free hydrogen
4.5 Atmospheres	24 %
... "	25 "
0. ... "	0 "
0.2 "	... "

Some ... to ... obtained at 26° and 350°
A most striking example is ...
... the works of ... Singer ..., in
which, ... shows that while oxygen
is absorbed by ba... ... at 450°,
barium dioxide being formed, yet ...
the pressure be there ... and all the
oxygen or ... ... of oxygen ...

off again, in other words, by means of a reduction, pressure we have effected this transformation, $BaO_2 \rightarrow BaO + O$.

<u>Power of Increased Pressure to Retard Decomposition.</u>

One of the most classic pieces of work of this nature is the work of Debray on the dissociation of Calcium carbonate, which was confirmed a little later by Mcc Gold... Both of these work... that f... g ... temperat... ... 750°... ... position ... ... definite pressure limit, ... Calcium carbonate be heated to 460° in a closed vessel Carbon dioxide ... is given off until a certain pressure within the vessel is reached, and then the decomposition will cease and...

...in a sufficient ... and calcium carbonate may be melted without decomposition.

Isambert has found the same to hold true for barium carbonate, and indeed the same author has studied the limiting pressures of about a dozen of the compounds of ammonia with the metallic halides (usually considered ... ular) and these conduct themselves exactly like calcium carbonate.

Some progress has been made in the study of inorganic salts containing water of crystallization, by a study of the vapor tension of these compounds. Quite suggestive ... have been obtained by Debray[2] from a study of the limiting pressures at different temperatures. Case most striking is a ...

Temperature	Limit Pressure	Temperature	Limit Pressure
12.3°	7.4 mm	31.5°	30.2 mm
16.3°	9.9 "	36.4° (salt melts)	39.5 "
20.1°	14.1 "	72.0°	80 "
24.9°	18.2 "		

Wiedemann[1] has studied the vapor tension of the following compounds: $MgSO_4 \cdot 7H_2O$, $ZnSO_4 \cdot 7H_2O$, $CoSO_4 \cdot 7H_2O$, $NiSO_4 \cdot 7H_2O$ and $FeSO_4 \cdot 7H_2O$. For each temperature a limiting pressure was observed, the results being quite similar to those obtained by Debray with calcium carbonate, except in those cases when the salt is capable of forming more than one hydrate.

True this may be in way at beginning, yet it is a beginning

[1] Ann. der Phys. Pogg. (1874) 474 and Jour. frak. Chem. [2]7, 338

toward the solution of the mystery involved in the constitution of these
compounds.

Effect of an Excess of one constituent on the Dissociation of Compounds.

Having observed the decomposing effect
of reduced press. and the opposite
effect of increased pressure, let us notice one case which shows the effect of an excess of one constituent; in this
way the partial pressure of the constituent present in smallest amount is
reduced, equivalent to an absolute reduction of pressure. A case in point is the
work of Lemoine on the dissociation of hydriodic acid, but one of greater histori-
cal interest is that of a... ium
carbamate. As was conclusively shown

by the work of Naumann, ...
he confirms the much earlier work
of ...... and of .... Ammonium carbamate is converted
into vapor it undergoes decomposition
into ammonia and carbon dioxide,
the amount of dissociation varying
with the temperature. Working at reduced pressures the results are as follows:
dissociation is greater in a vacuum
than in the presence of one of the constituents, and the greater the pressure
of this constituent, temperature remaining constant, the less the dissociation.
For each temperature there is a limiting pressure and the limiting pressure
is greater for a given temperature if
there is an excess of ...... than
than when ammonia is in excess

I mention this case for in all the decompositions made by me there was an excess of Carbon monoxide.

<u>Effect of a Solvent on Dissociation Tension</u>

Finally let us notice the effect of a solvent on the dissociation tension of a compound, the work of Greutur furnishes an excellent example. He finds that firm n.! dry acid sodium carbonate is not perceptibly dissociated below 25°C, in a vacuum. However it is very slightly dissociated in a vacuum of 10-20 mm at 25-30°C. For example

Original weight $NaHCO_3$ 4.4977 g
Weight after 48 hours <u>4.4948</u>
0.0029

If the temperature be now raised to 110-115°C dissociation is almost complete.

The same reduced press.ⁿ being sustained.

Original weight $NaHCO_3$		1.2166 gr.
Weight after 2 hours – 100° in a vacuum		1.1467
" " 8 " " "		0.9900
" " 20 " 100–110° "		0.7767
" " 48 " 100–116° "		0.7755
Loss in weight		= 0.4411
Loss required by $2NaHCO_3 = Na_2CO_3 + H_2O + CO_2$		= 0.4478

The presence of water causes an extreme difference in conduct. When 5 grams of acid sodium carbonate were dissolved in 20 grams water and the solution placed in a vacuum of 300–400 mm. at 25–30°, at the end of four days the salt was dry; the mixture contained about 18.7% of $NaHCO_3$ and 81.3% Na_2CO_3. The amount of water present effects very markedly the rate of decomposition; if the salt is only moistened the decompos.ⁿ tion is practically nothing being merely 2 to 3% instead of 81.3 when 20 grams of water are used.

The effect of water on the dissociation of acid potassium carbonate is exactly similar to that exerted on the sodium salt. Also Berthelot and André find[1] that dry acid ammonium carbonate suffers almost no dissociation, but in the presence of water or water vapor the tension is very markedly increased.

Let us consider again now the behavior of dicarbonyl cuprous chloride under reduced pressure and also when it is submitted to increased temperature.

As regards the ~~the~~ behavior under reduced pressure: but in... different

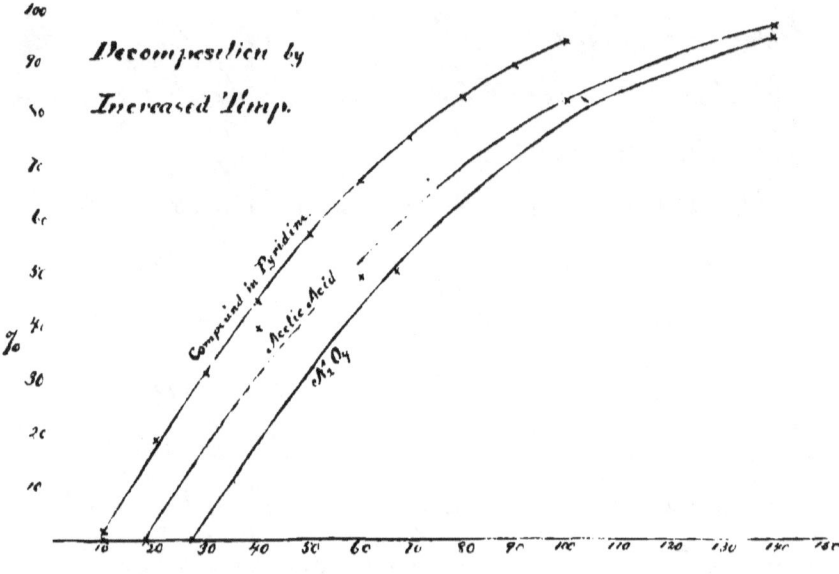

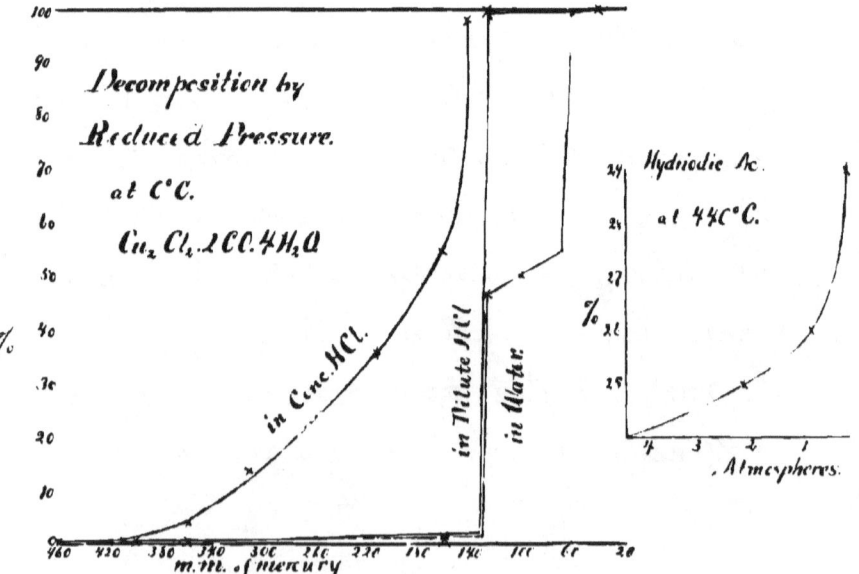

ways it conducts itself quite differently. In that instance in which carbon monoxide is absorbed with most difficulty (use of bur water), it is most difficultly given off by a reduction of pressure, and in that instance in which it is most easily absorbed (by use of hydrochloric acid), it is most easily given off. The compound made with dilute hydrochloric acid decomposes almost entirely at about 130 ^{mm} pressure. The break observed in the decomposition of the compound made with water is probably somewhat exaggerated in the curve (page 1336), though there undoubtedly is something of a break, yet it extends only through a very short range of pressure — from about 130 to 160 ^{mm} so that a very slight error in observation would lead to quite different results. I hardly think this slight irregularity, extending through only 30 mm, a sufficient reason for concluding that in the compound made with water, when we seemingly have a splitting off of one half of the carbon monoxide and then the rest, that of the two molecules of carbon monoxide in the compound, one molecule is united in a manner different from the other.

When concentrated hydrochloric acid is used, the decomposition is of quite a different kind, due most probably to the fact that in this instance a portion of the compound is in solution. Just how this could bring about such a change in conduct is of course unknown, but that the presence of a solvent does have a marked effect was shown in the work already cited on the dissociation of the acid carbonates of the alkalies. The dry materials show almost no tendency to decompose at the ordinary temperature, increasing little by little to about 100° where it is entirely decomposed within a few degrees; just as decarbonyl nitrous chloride decomposes in dilute acid only very slowly until a pressure of 135 mm is reached and there it is entirely decomposed with very little variation in pressure.

However the acid carbonates when moist dissociate slowly at the ordinary temperature, showing the great effect produced by the solvent

Further Dr. Rehschmidt has pointed out that carbon monoxide which has been absorbed by an ammoniacal cuprous chloride solution can be driven out by passing through any the solution an indifferent gas as nitrogen. On the other hand, I have found that when nitrogen or hydrogen is passed for eight hours through water in which crystals of dicarbonyl cuprous chloride are suspended, the crystals show not the least sign of alteration, provided the temperature is low; of course if the temperature is slightly raised, decomposition will begin.

As regards the decomposition, by increase of temperature, of the compound formed in pyridine by the absorption of carbon monoxide by cuprous salts

¹⁾ loc. cit.

oxide, no further comment is necessary. It is strictly analogous to the similar decomposition of compounds which must be accepted as atomic. The truth of this is strikingly shown by a comparison of the curves showing the decomposition of this compound and the decomposition of hydriodic acid, as was shown by the work of "Irvine". The analogous decompositions of acetic acid and nitrogen peroxide are also to be easily recognized.

The decomposition of decarbonyl or nitrous chloride and the exactly similar decomposition of hydriodic acid by nascent hydrogen, when the former is in concentrated hydrochloric acid, is so well shown by the curves that nothing further need be said.

summing up all the properties of the carbonyl cuprous chloride I can find nothing which could lead one to consider it as anything else than a true atomic compound. In definiteness of crystal form, constancy of composi- and in behavior under temperature and pressure changes it conducts itself like a true chemical individual.

The decomposition of a compound by a reduction in pressure is not unknown, as I have attempted to show, and yet it is unusual I confess. Our conditions of observation are so extremely limited, that we are inclined to believe impossible that which is not customary in our these limited conditions. But are not chemical compounds, as far as both their physical and chemical properties s concerned the result of their rearrangement? Do we exclude mercury from the list of metals simply because it exists in the liquid form at the temperature in which we happen to live? Indeed do we not look upon hydrogen, if not as a metal, yet very closely allied to the metals?

Further, do we consider nitrogen trichloride as being something different from a true compound because under our conditions its parts are so balanced that the least external disturbance destroys this equilibrium and makes its existence impossible? Not at all; the unusual is not the impossible and yet they are sometimes confused.

Confine ourselves no longer to such

narrow limits of temperature and pressure, give us other conditions and we shall have other compounds. At the temperature of the sun we should have no solid compounds— indeed very few compounds at all; instead of lakes and oceans we should have great volumes of oxygen and hydrogen. At the temperature of absolute zero, we should have no gases and certainly many of our compounds would be considered elements. Indeed our compounds are just on the border-land— stable enough to exist yet not so stable as to withstand our means of decomposition, and it is this very instability that enables us to class them as compounds rather than as elements.

Can we lay down certain exam

ments, cannot yet hard and fast, and demand that all compounds come up to these? Laws, as we know there are but our interpretations of natural phenomena. True nature is constant but our knowledge is incomplete and our judgement often false. With new facts before us we must modify and extend our ideas else we bind our hands with artificialities and blind our eyes to truths yet to unveil.

The distinction drawn by Kekule between "molecular" and "atomic" compounds cannot longer be held, any more than the conception which gave birth to the distinction — constant valency of the elements. I do not wish to imply that all compounds are of the same chemical order, but I do wish to protest against the use of the term "molecular compound" in the sense that Kekulé used it. It not only fails to tell us anything, but on the other hand causes us frequently to imagine that we have gotten some insight into a compound, eases the inquiring mind and so does positive injury.

It will be noticeable to all that I have written the formula $CuCl_2 \cdot 2CuO \cdot H_2O$ just as Kekulé would have written it; and indeed very similar to the old formulae of Berzelius, which conception seems the remote ancestor of the molecular idea. By those dots I do not mean to imply that between Cu_2 and Cl_2 there is one kind of attraction and between Cu_2Cl_2 and

Then = another, this is to represent not
a state of thing in the compound but
rather a state of difference in ourselves.
I think we can say that in the compound we have the elements in the proportion Cu_2Cl_2 and $2CO$ and $4H_2O$ and
report that we say nothing; indeed
it might be better to write the formula
$Cu_2Cl_2 C_2 H_8 O_6$

If when we speak of a compound
as molecular we mean simply that
we do not know how the various
parts are arranged as regards one another, well and good; but when we
speak of ammonium chloride, for example, as a molecular compound
and mean by this that the internal forces between the atoms NH_3
and HCl are of different kinds, when

we base our conceptions of the chemical relations of a compound in the usual state on an instability in the gaseous state; and worse than all, when we try to explain chemical phenomena by a distribution of names, then I do object to the term and believe that progress is hindered by such a usage.

Thirty years ago a term was introduced into the chemical language; no doubt it did some good, ~~introduced~~ as it directed the attention of chemists to a certain class of compounds, and caused them to inquire into the nature of the relations existing in them. A separate class was formed, but one by one these molecular compounds have been shown to be true chemical compounds until there seem to be only two classes of compounds behind which the upholders of Kekulé's conception of molecular compounds can entrench themselves. I refer to the compounds containing "water of crystallization" and the so-called double halides. Thanks to certain broad-minded conceptions, compounds of the latter class are fast being given their proper chemical status, and in a few notable instances those of the former class have received the same treatment. I refer especially to mesoxalic acid and chloral hydrate.

We see at once that real advancement has been made along these lines, and let us hope that the future may give us an insight far deeper than we now possess, knowing that, when these territories, once so dark, shall have been not only explored but taken full possession of by chemists, ~~that~~ larger lands will attract our attention, deeper problems will confront us; but with the new implement in our hands, which each succeeding year brings to us, and taught by the lives of those who have gone before, aided by a casting off of all useless armour and buoyed up and strengthened by the victories of the past, in shall attack the new problems ~~shall~~ with a renewed vigor and thus the progress of thought

Inorganic Compounds Containing Carbon Monoxide

Since the conduct of dicarbonyl cuprous chloride may seem to present many peculiarities it may be well to note briefly the other inorganic compounds containing carbon monoxide and to see if we can detect any peculiarities belonging to them as a class.

$COPtCl_2$ Schützenberger: Ann chim phys 1-4, 15, 10

and ibid. ,41 2, 350

$(CO)_2 Pt' Cl_2$ "

$(CO)_3 Pt Cl_2$ "

$(CO) Pt NH_2.2,t$ "   Bull. soc. p. chim [2] 14, 23

and ibid. [2] 14, 7..

$CO\ Pt\ K_2 \cdot 2HCl$ (R=primary base)   Compt. rend $70, 1287$

$CO_2\ Pt(NH_2)_2 \cdot 2HCl$ etc.   "

$CO\ PtCl_2 \cdot C_5H_5N \cdot HCl$ Foerster   Ber. $24, 2424$

$CO\ PtCl_2 \cdot C_9H_7N \cdot HCl$ etc.   "

$CO\ Pt\ Br_2$ "

$CO\ Pt\ I_2$ "

$CO\ Pt\ Br_2 \cdot C_5H_5N \cdot HBr$ etc.   "

$CO\ Pt\ CNS \cdot KCNS$ "

$CO\ Pt\ S$ "

$CO\ PtO$ "

$(CO)_2 Pt\ Cl_6$ or $PtCl_2 \cdot 2(COCl_2)$ Pullinger J. Chem. Soc. $59, 598$
                              and   Ber. $24, 2291$

$CO\ PdCl_2$ Fink   Compt. rend. $126, 646$

$(CO)_2 PdCl_2$ "

$(CO)_3 (PdCl_2)_2$ "

$Ni(CO)_4$ Mond, Langer and Quincke   Jour. Chem. Soc. $57, 749$
                              and   Chem. News $64, 108$
                              Brit. phys. Chem. $8, 150$

$Fe_2(CO)_5$ Mond and Langer   Jour. Chem. Soc. $59, 604 + 1090$

$Ir_2(CO)_7$ Mond and Langer Jour. Chem. Soc. 57, 1090
$Ir_3Fe(CO)(CN)_5$ Muller Compt. rend. 104, 994
  and Ann. chim. phy. 5(6)4, 94
$Pt_4(CO)_7$ (trepa. unknown) Hartrek & Lenge. Zeit. anorg. Chem. 16, 50
$(Pd)_4(CO)_7$ " " "
$COCl_2$
COS

To review now briefly some of the properties of these compounds. All of them are more or less unstable and on decomposing tend to give off carbon monoxide as such, unless an oxidizing agent be present, when all or a part of the carbon monoxide appears as carbon dioxide.

Of the previous compounds CO_2, COS and $Ni(CO)_4$, COS breaks down at a dull red heat into CO and S and at about 200° $Ni(CO)_4$ is decomposed into Ni and $(CO)_4$. The other compounds with the exception of $Fe(CO)_5$, undergo decomposition when converted into vapor, the carbon monoxide usually appearing as such and indeed $Fe(CO)_5$ is broken down at 200–350°.

The compounds $CO\cdot PtCl_2$, $(CO)_2 PtCl_2$ and $(CO)_3\cdot(PtCl_2)_2$ as well as $CO\cdot PdCl_2$, $(CO)_2 PdCl_2$ and $(CO)_3(PdCl_2)_2$ can all be converted one into another by changing the temperature and the proportion of carbon monoxide present, and all of them are decomposed by water in the cold. In the case of $CO\cdot PtCl_2$, and also $CO\cdot PdCl_2$, when decomposed by water all the carbon monoxide appears as the dioxide, due probably to the intermediate formation of carbonyl chloride which

... is acted on by the water. When an excess of carbon monoxide is present as in $(CO)_2 PtCl_2$, just one half of the carbon monoxide suffers oxidation. These decompositions are

$CO PtCl_2 + H_2O = Pt + CO_2 + 2HCl$
$(CO)_2 PtCl_2 + H_2O = Pt + CO_2 + 2HCl + CO$
$(CO)_3 (PtCl_2)_2 + 2H_2O = 2Pt + 2CO_2 + 4HCl + CO$

The corresponding palladium compounds are decomposed in an exactly similar manner.

The compounds $CO Pt(NH_2)_2 . 2HCl$, $(CO)_2 Pt(NH_2)_2 . 2HCl$, $CO Pt I_2$, $CO Pt Br_2$ etc. are also decomposed by water. When $CO Pt Br_2$ is heated in an atmosphere of carbon dioxide it is decomposed thus:
$CO Pt Br_2 \rightarrow CO + Pt Br_2$.

When a solution of $CO Pt CNS . KCNS$ is simply evaporated carbon monoxide is not firm. The compound $COPtS$ is so unstable that it decomposes in a vacuum.

All of the compounds of the types $COPtCl_2$, $COPt(R)_2·2HCl$, $CO_2Pt(R)_2·2HCl$ (R = residue of ammonia or any primary base), $COPtCNS$ etc. are decomposed by the action of potassium cyanide, carbon monoxide appearing as such in every case.

Even from this exceedingly brief review I am sure we can detect great similarity between dicarbonyl cuprous chloride and the above compounds, a similarity so great that one could not fail to classify it with them as a true chemical compound.

On account of the similarity in composition between $CO_2·Cu_2Cl_2$ and an! The compound CO_2, H_2, I have given to the former compound. The name which I have used all through this paper — dicarbonyl cuprous chloride — being in accord with the name given to the latter compound by Schützenberger.

Conclusions.

I. Carbon monoxide is not absorbed by dry cuprous chloride. It is absorbed by cuprous chloride when this substance is in solution in concentrated or dilute hydrochloric acid or even when suspended in water.

II. As a result of this absorption a compound of definite composition is formed, identical in all cases. ~~which~~ This is to be represented by the formula $Cu_2 Cl_2 . 2CO . 4H_2O$.

III. This compound conducts itself like a true "atomic" compound. It is decomposed by a reduction of pressure or by an elevation of temperature, but these decompositions are exactly like the decompositions of other substances, the true "atomic" nature of which cannot be doubted.

IV. Crystals of dicarbonyl cuprous chloride are unaltered at 0°C by the action of hydrogen or nitrogen. They are immediately decomposed, however, at this temperature, by oxygen or chlorine. It seems that it is the cuprous ion which is the sensitive constituent under the influence of these reagents.

V The Carbon monoxide shows no marked increase in chemical activity at the moment of libration from the compound.

VI From a comparison of decarbonyl cuprous chloride with the other inorganic compounds containing carbon monoxide, it is found to conduct itself in a manner similar to them, and deserves to be ranked with them as a true chemical compound.

Biographical.

William App Jones was born in Hillsboro, N.C., Nov. 24th 1873. His early training was received in the private schools of that place.

In 1886 he entered Moravian Falls Academy, and after three years of preparatory study he entered Wake Forest College (N.C.), from which institution he received the degree of Bachelor of Arts in 1893. After being in business for a year he returned to his Alma Mater, where he spent a year in the study of chemistry.

In 1895 he entered the Johns Hopkins University where he has since devoted himself to the study of chemistry, geology and mineralogy.

In the spring of 189? he was

appointed a Fellow in Chemistry the present year and since Jan. he has acted as lecture assistant to Prof. Remsen.

www.ingramcontent.com/pod-product-compliance
Lightning Source LLC
Chambersburg PA
CBHW030254170426
43202CB00009B/736